ANIMAL BABIES

BABY EAGLES

by Jenna Grodzicki

Consultant: Beth Gambro
Reading Specialist, Yorkville, Illinois

BEARPORT PUBLISHING

Minneapolis, Minnesota

Teaching Tips

Before Reading

- Look at the cover of the book. Discuss the picture and the title.
- Ask readers to brainstorm a list of what they already know about eagles. What can they expect to see in the book?
- Go on a picture walk, looking through the pictures to discuss vocabulary and make predictions about the text.

During Reading

- Read for purpose. Encourage readers to think about eagles and the animal life cycle as they are reading.
- If readers encounter an unknown word, ask them to look at the sounds in the word. Then, ask them to look at the rest of the page. Are there any clues to help them understand?

After Reading

- Encourage readers to pick a buddy and reread the book together.
- Ask readers to name three things eagles do between the time they hatch and the time they are ready to have babies. Go back and find the pages that tell about these things.
- Ask readers to write or draw something that they learned about eagles.

Credits:
Cover, © dpa picture alliance archive/ Alamy; 3, © Tanya_ Goncharova/ Shutterstock; 5, © orin/ Shutterstock; 7, © predrag1/ iStock; 8, © Jon C. Beverly/ Shutterstock; 9, © Prostock-studio/ Shutterstock; 9, © Shchipkova Elena/ Shutterstock; 10, © xpixel/ Shutterstock; 10,11, © BirdImages/ iStock; 12, © Eugene Lagana/ Shutterstock; 14,15, © Tracy Kerestesh/ Shutterstock; 15, © watin/ Shutterstock; 17, © pchoui/ iStock; 18, © Al kearney/ Shutterstock; 20, © watin/ Shutterstock; 20,21, © slowmotiongli/ iStock; 22, © predrag1/ Getty; 23, © orin/ Shutterstock; 23, © davidhoffmann photography/ Shutterstock; 23, © David Gowans / Alamy; 23, © BrianEKushner/ iStock; 23, © matimix/ Shutterstock.

Library of Congress Cataloging-in-Publication Data

Names: Grodzicki, Jenna, 1979- author.
Title: Baby eagles / by Jenna Grodzicki.
Description: Bearcub books. | Minneapolis, Minnesota : Bearport Publishing Company, [2022] | Series: Animal babies | Includes bibliographical references and index.
Identifiers: LCCN 2021026714 (print) | LCCN 2021026715 (ebook) | ISBN 9781636913551 (library binding) | ISBN 9781636913629 (paperback) | ISBN 9781636913698 (ebook)
Subjects: LCSH: Eagles--Infancy--Juvenile literature.
Classification: LCC QL696.F32 G76 2022 (print) | LCC QL696.F32 (ebook) | DDC 598.9/421392--dc23
LC record available at https://lccn.loc.gov/2021026714
LC ebook record available at https://lccn.loc.gov/2021026715

For more information, write to Bearport Publishing, 5357 Penn Avenue South, Minneapolis, MN 55419. Printed in the United States of America.

Contents

It's a Baby Eagle!

Crack!

A baby eagle is **hatching**.

It takes all day to break out of its egg.

A baby eagle is called an **eaglet**.

Its **feathers** are soft and fuzzy.

Some eaglets are gray.

Others are white.

Say eaglet like EE-glet

The baby is the size of an adult human's hand.

It stays in a nest with its mother.

She keeps the eaglet warm.

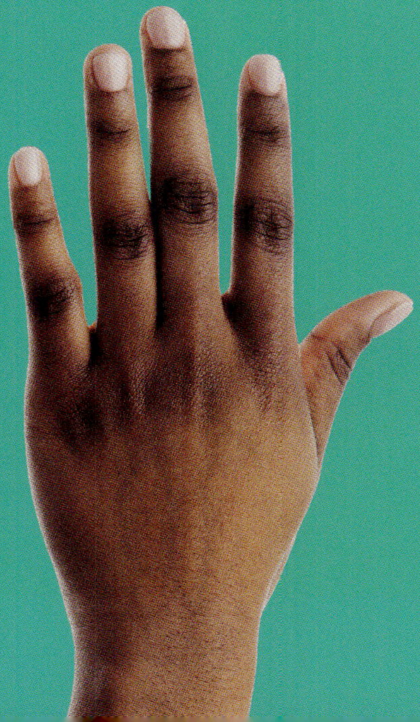

The baby's father brings food to the nest.

Then, the mother feeds the hungry baby.

Gulp!

The eaglet grows fast.

In six weeks, it is almost as big as its parents.

It starts to grow new feathers, too.

The eaglet flaps its wings.

It hops around the nest.

Then, it jumps to a nearby branch.

When the eaglet is 10 weeks old, it starts flying.

At first, it cannot fly very far.

So, the young bird **practices** every day.

Soon, the eaglet can fly well.

It starts **hunting** for food.

But it still goes back to the nest to eat and rest.

A few months later, the young eagle leaves the nest.

It is ready to live alone.

In five years, the eagle can have its own babies.

The Baby's Body

Glossary

eaglet a baby eagle

feathers the things that cover a bird's body

hatching coming out of an egg

hunting looking for another animal to eat

practices does something many times to get better at it

Index

Read More

Gaertner, Meg. *Eaglets (Animal Babies)*. Lake Elmo, MN: Focus Readers, 2020.

Schuh, Mari. *The Bald Eagle (Blastoff! Readers: Symbols of American Freedom)*. Minneapolis: Bellwether Media, 2019.

Learn More Online

1. Go to **www.factsurfer.com** or scan the QR code below.
2. Enter "**Baby Eagles**" into the search box.
3. Click on the cover of this book to see a list of websites.

About the Author

Jenna Grodzicki loves to read and go to the beach with her husband and two children.